Lean Manufacturing
—CONTINUOUS IMPROVEMENT—

Common Sense, Not Rocket Science

RANDALL L. KIES II

Copyright © 2021 Randall L. Kies II
All rights reserved
First Edition

Fulton Books, Inc.
Meadville, PA

Published by Fulton Books 2021

ISBN 978-1-63860-893-6 (paperback)
ISBN 978-1-63860-894-3 (digital)

Printed in the United States of America

To my wife, Angelina, who has been an integral part of this writing. She is the voice of the employees who have been on the receiving end of poor training and terrible application of the tools within Lean/CI. She spent thirty years on the factory floor, and she will be the first to tell anyone that Lean/CI is not rocket science. It is clearly common sense and should be treated that way.

Contents

Preface 7
Setting the Record Straight 11
Head Count or Profit 19
Example 1 22
Example 2 26
Lean/CI in the Classroom 30
Short Story 33
Easy Money 35
Example 3 37
Timing Is Everything 41
Reliability and Availability 44
Who? What? Where? When? Why? 52
Task Lists 55
Final Thoughts 59

Preface

How much money have we saved? How many people were laid off? One manager may argue that there are no savings generated from Lean Manufacturing (Lean) activity unless you are reducing head count. Another manager may think the savings contribute to additional sales and profit, streamlined operations, and additional investment in the company's future. There is also the human element where employees may enjoy a boost in morale. Which is it? Who's right?

Where do you stand with regards to Lean thinking and continuous improvement (CI) implementation? Is it worth the investment? What do you think with regards to savings calculations? Do we have to reduce head count to claim savings? Can we use additional sales, improved quality and safety, and streamlined operations to quantify savings and reinvest in the future?

After thirty-five years of learning, teaching, and practicing Lean/CI methodology, it seems as if savings is the item that creates the most debate. It is obvious that reducing head count is a definite cost saving for a company. That is considered a "gimme" as far as I'm concerned because it is the easiest savings to calculate and achieve with very little effort. Honestly, it is generally understood that head count changes with the tides of sales. More sales equal more people and vice versa.

There has always been the thought that others will pick up the slack in order to make up for the difference when reducing head count. Historically, this is true, but people are getting tired of working harder and not smarter. Management wants more for less, but the only way to get there is to lay employees off.

Sadly, this thinking is archaic, and our companies are falling back in their ability to compete in a global market as a result. Companies are also missing the opportunity to reinvest in growth to become even more competitive. There are savings to be had, and head count reduction is not the only way to quantify improvement activity.

What about the other stuff? Are those savings as well? Here is where the argument always begins, and common sense takes a back seat to the egos at the management level. Some managers are telling me that we cannot claim savings on quality, efficiency, or additional capacity improvements.

Really? Why would you believe that? Prove it to me. Please explain why I am wrong. The response every time is crickets. Although I have not challenged my managers in this fashion, the scenario comes to mind every time the subject comes up. The idea of saying no because you can no longer holds any value to me.

My thought is the managers that argue against anything less than head count are sadly misinformed. They cannot see the forest through the trees because they are expecting a "home run" for every project that is completed. They want to see a change in head count, and anything less is unacceptable.

Success is about how the manager looks and not about the achievement of the team. Honestly, these people are out of touch with reality and have completely undermined the real purpose of Lean/CI. The practice of the craft is treated like rocket science, and common sense is thrown out the window.

The goal of this writing is not to argue the types of savings that can be generated. Instead, the purpose will be to share the real calculations that add up to real money without any argument. It will also focus on the common-sense application of Lean/CI tools in place of old-fashioned, status-based, uninformed decision-making.

These types of decisions have cost companies a lot of money. Some companies don't even realize how much opportunity is missed by allowing managers to make bad decisions. The calculations for the improvements referenced within were developed during my tenure as a manufacturing/Lean/CI engineer. They have also been blessed by the finance department of two large manufacturing companies in the United States.

The third aspect of this writing will be to encourage the engagement of employees in the improvement process. Top-down Lean/CI doesn't work. The house of Lean/CI clearly shows that the foundation must be laid first. However, many Lean/CI practitioners are too excited to show everyone how their vast knowledge is the only way to transform a company. I beg to differ. And for many reasons.

The final aspect of this writing will be to make the case to move Lean/CI into the K–12 classroom to teach the basics prior to high school graduation. Science, technology, engineering, and math (STEM) have been identified to improve how we teach our children and improve our ability to compete on a global scale. Teaching Lean/CI principles in high school will open up opportunities to customize training after graduation.

The idea is each company can build on the basics and create processes that work specifically for the company. In doing so, each company can be more competitive because their process is confidential, not the application of basic Lean/CI methodologies.

It only makes sense that we use standard Lean/CI applications to create a competitive advantage for ourselves; however, it makes no sense to continue to bastardize and plagiarize Taiichi Ohno's work. There are far too many self-proclaimed experts that don't know how to apply basic principles, and this is one area that needs vast improvement.

Lean/CI implementation must be built from the ground level in order to establish a good foundation. People must be trained in the basics before moving on to more difficult improvements. The combination of class time and working on the floor is critical to how people learn and retain Lean/CI. Finally, the cross section of the team, not one individual in particular, leads to success.

Setting the Record Straight

The very first order of business is to set the record straight for the reader. Taiichi Ohno developed the Toyota Production System (TPS) shortly after World War II. There are hundreds of company's "business systems" that have adopted the teachings of Mr. Ohno but then changed a few words and called it their own. My opinion is this is acceptable as long as you recognize Taiichi Ohno as the originator. The practice is called proper referencing of source information.

I worked on the development of continuous improvement programs at several companies where the Toyota Production System was used as a guide. This is an advantage when training people because the methodology is global. Standardization across the Lean/CI methods lends itself to better retention and application of Lean/CI tools. In short, the basic training from another company should fit in with the training at your company because both are based on TPS. The basics are the same for everyone and essential no matter where you work.

This would be a novel idea, but some companies can't standardize work between disciplines, departments, and plants. It's as if every plant in the operation has a different plan for Lean/CI implementation, and they are all based on the beliefs of individual managers. There seems to be zero connection with or without a corporate continuous improvement team.

Recently, I had interesting conversations with four different managers at the same company. One of them took sole credit "for thirty years of program development of Lean/CI implementation of their business system." The other three managers had a strict belief that savings are only generated by a reduction in head count, and anything less doesn't exist. It seemed like a home run or nothing would be the only possible outcome of any continuous improvement event.

The conversations were extremely awkward because there was no reference to TPS or Mr. Ohno. It also felt like there was zero ability for any of these people to get out of the box and think differently. They were set in their thinking, and they all outranked me, so there was no argument that would convince them that I was right. This is the exact reason for writing this book. After thirty-five years of disrespect, it's time to set the record straight.

Over time, I have given up trying to reason, argue, prove wrong, and convince upper management that savings can be identified in many places other than head count. Instead, I worked with the finance departments and found out how to identify savings in several areas. I did not ask for permission then, and I will not apologize for the improvements my teams were able to make.

The finance people and I found savings in areas such as machining, part kitting, product assembly, preventive maintenance, and overall availability; streamlining operations and workflow were also identified. It doesn't take a rocket scientist to realize that it is possible to do more with less, and redeploying workers to add value in another area is an actual cost saving.

First and foremost, anything that can be done to improve one or all of a company's key deliverables should be considered a win. Head count will vary based on the level of sales, so focus on operational improvement wherever possible. The goal is to stabilize the operation in order to weather the ebb and flow of sales. I consider this the bottom-line approach, and it has served me well.

A former supervisor referred to this as having the gas pedal pushed three-fourths of the way down and then having the ability to adjust to meet the changes in production. It is a simple visual that explains the complex workings of operations—well said in my book.

It is also important to evaluate quality, safety, efficiency, availability of assets, and standardized work instructions. You have to get out of the box, identify the obvious, and make some improvements.

It really is that simple. Additionally, keep paperwork to a minimum, but make sure to capture the "before and after" of the improvement metrics.

There are areas that are easy to improve such as product flow, component storage, area layout, 5S, and visual management. The identification of "incoming" and "outgoing" products reduces the need to interrupt people who are busy doing their job. Organizing tools so they are always in the same spot irritates some people, but the fact is it should take less than ten seconds to obtain or return tools to their proper location.

Another point that must be made is that Lean manufacturing and continuous improvement are a global phenomenon. Lean/CI does not succeed because of one person or any false claims of greatness. It doesn't succeed because one person can argue every detail of the company's business system. Finally, Lean/CI is a matter of contributing to the solution, not continuing to support the problem.

Lean/CI thrives on teamwork and the application of the tools used to make improvements. Lean/CI is successful through training the basics in the classroom and applying that training on the manufacturing floor. The purpose of the Lean practitioner is to teach the appropriate tools at the proper times to achieve maximum impact on the items being improved. Anything more than that is simply fluff and propaganda.

Here is the point. There is an old saying that claims "knowledge is power." The thinking behind the statement is horribly outdated. The best way to have an advantage over coworkers and other employees was to gain knowledge in a trade or a specialty. Additionally, the individual would hoard the information for personal gain. Management views this as leadership because the individual stands out from the crowd.

However, the individual begins to separate from the group, and the knowledge is used against coworkers as a competitive advantage. The behavior was seen as forward-thinking by the management; however, coworkers lost respect and confidence for these individuals as the powerful knowledge went to their heads.

The need to dictate every action necessary to produce a dimensionally acceptable part has not changed. Lean/CI offers several tools to ensure accountability and quality output without the use of intimidation and scare tactics. Honestly, the days of one person having all of the power are over, and this behavior is completely unacceptable and outdated.

Fast-forward to the present time, and ask yourself if this behavior is still acceptable. Many Lean/CI practitioners will answer yes because they feel their knowledge is going to make the difference between success and failure. I beg to differ because one person will not make the difference. The team makes the difference, and I am throwing the bullshit card.

I came up with a phrase claiming "knowledge is not power; common knowledge is power" to offset the feeling of dictatorship in an improvement setting. It served me well because it made everyone in the room peers, including myself. It also lowered the anxiety of dealing with attitudes and egos. Frankly, there is no time for that, especially when the current event is packed with opportunities.

Additionally, I was able to tap into the knowledge of the entire team and not follow an agenda that serves someone else. I have used this technique to build successful teams for thirty-five years so it's easy to say it works well for everyone, not just the Lean/CI practitioner. The results are simply incredible when everyone on the team can work together.

The teams I worked with were able to overcome impossible odds to accomplish some amazing things. We worked together and pooled our thoughts instead of blindly following someone else. Personal integrity, respect for each other, and building on ideas were paramount; so maybe it's time for some to consider a different approach.

The most important part of team development is the cross section of people. Bring people on board with the expectation each individual is the "subject expert" in their chosen craft. Treat them with respect, and the team will benefit as a result. The point is to recognize and use the knowledge of each team member because one person cannot know everything.

Subject experts exist in almost every occupation that I have interacted with. A few examples would include design, quality, process, and manufacturing engineering. Other areas include maintenance, facility engineering, operators, assemblers, electricians, and material handlers. Depending on the size and structure of the company, management experts exist in accounting, finance, operations, and Lean/CI. This is not an all-inclusive list because it doesn't matter what the title is. Bring the right people to the table.

For example, it doesn't make sense for a team of assemblers to lay out a machining center. It does make sense to include one individual from the assembly because they are the next operation in the process. The bottom line is to get the right people at the table and tackle the difficult issues necessary for improvement. Leave ego and emotion at the door because everyone on the team needs to have equal input.

Unfortunately, many of the Lean/CI practitioners I have encountered focus on justifying their position and not on actually contributing to the overall success of the company. For whatever reason, it seems necessary to argue every gnat's-ass detail of a business system instead of engaging employees from the necessary areas to generate improvements. The problem with this thinking is everybody loses, especially the company.

The practitioner doesn't learn anything from feedback because he/she is not really listening to others and, by default, is not engaged in the improvement. They tend to speak *at* the audience and not *to* them. Flexibility is not part of the agenda, and interaction with the group is minimal at best. I'm not sure about the reader, but this is unacceptable in my world.

Unfortunately, I have to refer to my children's first-grade teacher to clarify my point of view. She made the following statement to a room full of six-year-olds, and I believe it fits this situation. She said on the very first day of class, "If your lips are moving, then you aren't listening." That was twenty-five years ago, and it still makes sense.

Let me make one thing clear at this point, and I will repeat this as often as necessary to get the point across. If someone requests documentation of something that adds no value to current or future project reference, then it is a waste of everyone's time and money.

Lean/CI is not about creating useless information and charts for someone else. It isn't about how much a practitioner can make people do things that make them look good. Lean/CI is all about creating value for the company, and the justification of your job is in the savings generated through improvements. If the reader takes one thing from this writing, it should be understood that a Lean/CI practitioner is expected to guide the team and not dictate the change.

In general, Lean/CI engineers are required to return three times their salary back to the company each year. If you earn $100k, then your goal is $300k. This can be difficult if you have to do this alone; however, you can tap the expertise around you. Instantly, it becomes much easier to meet and beat the goal. After all, the improvements have to be proven and sustained by the people that perform the work. It's simply common sense to include them from the start.

The key to each event is teaching, learning, and engagement. Rigid training will lead to boredom, and employees get nothing out of the event except lunch. Additionally, lack of interaction with the audience places a wall between the practitioner and everyone else. Once the wall is built, the results of the project will suffer dramatically.

There will be less interaction, less ideas, and less participation, which directly leads to project failure. Employees disengage and have no motivation to make someone else look good. It is counterproductive at best and does not empower people to suggest improvements. The point is to recognize the subject experts and help them help everyone else, which helps the company.

As I have already pointed out, Lean/CI is a global phenomenon. It can be seen in places such as Pizza Hut, Taco Bell, and the most technical companies in defense and aerospace. Lean/CI is universal, and our advantage comes from proper application and focus on improvements. The advantage is not in the level of education or number of certificates a Lean/CI practitioner has on the wall. The advantage comes with engaging people and the expertise they bring to the table.

It puzzles me every time a Lean/CI practitioner speaks to a group in a disrespectful or condescending way. The reason is this person is going to turn around and ask the people in the group for help. I cannot count the number of times that this happened or the number of people who were blatantly offended. Once this happens, the only reason people stick around is for the free lunch.

Employees have no incentive to engage in future projects because they were treated like they were stupid. Additionally, they clearly understand that the underlying goal is head count reduction. Their belief is the practitioner has already decided what's going to change, and the opinion of the team really doesn't matter.

Seriously, I have seen many projects fail because of this, but I was always outranked by the person that did it. If you aren't realizing savings on the improvements, then it's time to change your approach.

Over time, it's been important to note these behaviors for several reasons. The first reason is my belief that Lean/CI succeeds through engaging people. People react positively when they are given credit for their work. They also like to be heard when they speak.

One sidenote is people will try to contribute to success because they were part of the process of improvement. It's critical to the overall success to recognize subject experts and not shut down their contributions. My experience has shown that people will begin to offer ideas without solicitation.

Obviously, you can't implement everything at one time, so it is beneficial to create a list of improvements. Use the list to set priorities and combine items that make the largest impact when finished.

The last thing is that most people in manufacturing learn by doing and really do not want to sit in a class all day. It's been my experience that moving to the manufacturing floor to apply the training helps people retain the information better. The best way to describe this thinking is I have a lot of education and training, but the people on the floor have the expertise to make improvements. If I am able to combine my education and training with the skill of the employees, then anything is possible.

Let me finish this section by saying employees who seem apprehensive to participate in improvement events usually have a very good reason. When asked, the majority of the employees said it was because of the past behavior of a Lean/CI practitioner. There was little respect simply based on condescending remarks and behavior during a previous project.

It goes back to the belief that the improvements are based on one person's plan, not a solution that the team has come up with. Support for a project like this is gone before it even starts.

Lean/CI practitioners need to recognize the capabilities of the team through its diverse makeup. The goal is to get the group thinking together with purpose in order to improve something. There are several tools that can help improve the chances of success, but it all starts with treating people with respect. My preference is to get the right people together, create a plan of attack, and get to work as a team.

The final element to share with regard to Lean/CI in the workplace is the permission to have fun during the event. Every project I run includes the permission to have fun—without anyone being hurt in the process. Seriously, I have made prototype piston rings, built custom fire trucks, coated parts for aerospace, and many other things that could cause catastrophic outcomes upon failure. The truth is boring projects yield boring results.

Do not misinterpret the intention of allowing fun, because the results have been overwhelmingly positive. People tend to lower their inhibitions when allowed the opportunity to pitch in without criticism. If you look at a coworker and they smile, it's probably because they aren't bored and someone actually listened. Many people have improved their understanding of Lean/CI because we had fun implementing small and large improvements.

Feedback and comments from projects were always positive, and the results were always above the expected outcome. People were willing to cross-train others because there is a "best process" that covers the important steps.

The key to avoiding pushback from others is to engage everyone involved and show why the new process is better. If you develop a process and force people to comply, there will likely be a loss in key measurables such as productivity, efficiency, and quality.

Head Count or Profit

Changing gears, why do we focus on head count when there are clearly savings in other areas? Even though I refer to headcount as a gimme, the effect on employee morale and their willingness to participate in improvement activity has been devastated. Who would want to participate in an event that eliminates jobs? Why would anyone participate just to see coworkers laid off?

Turn the situation around, and ask people for help because it makes the company more competitive. We can keep employees to redeploy them to another area or dedicate them to different support activities. We do not have to reduce the head count to appreciate savings. The real value is added through the retention of our workforce because it allows the company to do more with the same resources, and nobody loses their job.

The commonsense approach to Lean/CI is meant to help people work smarter, not harder. Combine ideas that make the work life of an employee better. The result should be beneficial to the employee and the company so it's a win-win situation.

There are advantages to small companies such as improved cash flow and working capital for growth. Additional benefits come with selling and delivering more products. It is a winning combination to either way you apply the savings. My preference has always been to keep the money inside the company and use it for growth. The hardest part has been trying to convince management that this was a good idea.

The idea of moving away from head count starts with some basic principles. The first is to stop looking for a home run or a remarkable step. My experience has been managers are looking for something to impress their boss, and big-ticket items pave the way for their promotions. Unfortunately for them, the truth is one home run or stepped improvement is nothing more than an organized group of small improvements.

Allow me to expand on this thought. One project that I worked on was to relocate an assembly department to another plant to reduce the travel of manufactured components. Moving the department alone would contribute to a 54 percent reduction in component travel, and management was ecstatic. The justification for the project was approved without hesitation, and anything the team came up with was a bonus. The story gets better from there.

Our team was able to implement standard tools for electricians and assemblers, create standardized layouts for a bay-build process, and create point-of-use storage for each assembly bay. Additionally, we deployed a water spider for delivering parts to each bay in order to reduce the number of employees leaving the department at any given time. At the end of the day, component travel was reduced by 54 percent, and we improved and sustained our assembly time by a remarkable percentage.

The point is we took a good project and made it better by applying some basic principles of Lean/CI in a team setting. Just to drive the point a bit further, an older employee approached me at the start of the project and said it would be a miracle if we could simply move the department. I asked him to trust me, and he damn near died laughing.

There is an acronym that comes to mind as I share this story. Many years ago, someone referred to the "trust me" statement as a BOHICA. What in the world is a BOHICA? The employee shared that it meant "Bend over. Here it comes again." It was another line of crap in the eyes of the employees, and it seemed like I just committed a cardinal sin. Anyway, at least we know what BOHICA means.

Two months after we were in our new area, the same employee walked up to me to congratulate me for making it happen. He liked the layout, the tools, and the fact that he didn't have to wander around looking for parts anymore. The biggest compliment he gave me was we did this without reducing one employee. On the contrary, we redeployed people and realized efficiency improvements that allowed the company to deliver more products to the customer. We will return to this in "Example 3."

Let me take this a bit further by adding this employee became an advocate for improvement because I was able to convince him through action, not just words. Another thing this man said to me was how refreshing it was to see people work together—especially with an engineer. I took it as a compliment based on my experience at other companies.

The real secret with Lean/CI is to strive for perfection through using the proper tools at the proper time. There is no way to do everything at once, so the plan should be to incrementally improve the right things at the right time. This can be accomplished by training to the necessary material, engaging people, and focusing on the manufacturing floor.

Example 1

I laid out a two-year project to improve the availability and reliability of two large assets. The current availability of the machines was 35 percent and 50 percent respectively, and unplanned downtime for each was about 30 percent. Quality suffered because of poor machine performance when production did run, so scrap and rework were also at unacceptable levels. The spare parts inventory for the machines consisted of many obsolete parts, and parts that had no min/max in the system. Finally, there were no written procedures for any repairs that had to be done at required intervals.

The machines were in bad shape, and the team had very little historic information go on. The only way to find the history for the machines was to sort through the previous six months of maintenance history. The root cause of the entire situation was the neglect of proper maintenance over the life of both machines. There was never any time to take the machines down, so "run to fail" was the accepted culture.

The time it would take to repair either machine properly was one week from start to finish. Needless to say, the enthusiasm and support for taking either machine down lacked management approval. Let me make one thing very clear. If you do not repair machines at the proper time, they will dictate unplanned repairs at the most inconvenient time.

The situation with large assets is they make a lot of money when they are running properly. It is extremely difficult to convince management to bring down a functioning machine. The classic response is to "let it fail; we will fix it then." Unfortunately, the choice must be made for long-term success or short-term commitments that may lead to quality issues and further failures.

The goal of world class manufacturing (WCM) is 85 percent for overall equipment effectiveness (OEE), and there are three contributors to the goal. Machine availability is most important to me because machines do not make money unless they are running properly. More availability equals more parts. The question is whether or not the machines are producing good quality parts. What good is availability if you are producing scrap?

The second measure of OEE is the level of quality or the measurement of bad parts in a run. The point I always make is quality depends on how well the machines are running. It's great to see the machines running, but are they capable of producing a good part? The answer depends on how you look at the assets and the level of quality you expect to achieve.

The third measure of OEE is efficiency, which is how well the people are utilizing the available machinery and information to make good parts. There are other aspects such as work instructions, tool placement, and area layout that also apply here. The key to success is knowing what to do to improve throughput in each area and within the entire plant.

Long story short, the goal was to bring availability up to 90 percent for each machine. One machine was twenty years old and had lasted twice as long as expected. The other machine was less than five years old and was halfway through the expected life. The case for both machines was neglect, and there was no overnight fix.

Ultimately, the improvement in machine availability was worth several million dollars in sales and new product development. Over time, we were able to bring both machines into the 85 percent range of overall availability and sustained the improvement for as long as management respected the maintenance schedule. It must be understood by all stakeholders that deviation from the maintenance plan may eventually lead to disaster.

The overall project included standardization in machine design, creating a spare parts inventory, standard work, and a preventive maintenance plan. Success would depend on the team, and we had a huge mountain to climb.

The plan for the two years included several smaller projects. Each project had deliverables that were critical to the two-year plan. The overall goal was to build from each project and hit the goals along the timeline to maximize the effect of the improvement. The point is the deliverables drove the project, and they were backed up by the necessary data to justify our plan moving forward.

On the contrary, management wanted continuous regurgitation of the same data, daily electronic updates, and savings per day reported to the plant manager. It was a ridiculous request that added absolutely no value to the immediate needs of the team. Additionally, the underlying tone was how much additional crap I would do to impress management.

Here is where everything goes south for me with regards to creating and saving unnecessary information. I am elbow-deep in a project worth millions of dollars to the company, and management wants me to make charts and graphs to justify their existence. Honestly, this is a clear example of too many layers of management and too many people taking credit for someone else's work.

There was no need to jump through hoops along the way because the team knew what had to be done. We didn't need more data to prove we were on the right track because the production numbers proved it. My feeling was we were beating a dead horse with the data, and the paperwork that came with it held no value to moving forward.

Don't get me wrong as far as Lean/CI tools are concerned. We used five whys and A3 problem-solving to get to the root cause. We also employed Pareto charts of the breakdowns and determined the priority of repairs. However, we didn't feel it was necessary to backtrack and redigest all of the information we already processed.

We could have, and we did, draw a line in the sand to identify what, why, and how often the machines were having unplanned failures. It was our baseline evaluation. Here again, an argument ensued because we were being asked to document more than we were solving. It made someone else look good to their boss, but it slowed the progress of the project significantly. The only thing I can say beyond that is the pushback was incredible, but I didn't have time to make pretty charts and graphs for someone else.

The point is there is no requirement to collect a ton of data and analyze it constantly as it applies to production. This is the basis of process control and can ruin baseline data that has already been analyzed. There is no value added once a project has been laid out and the process of improvement has begun. Production data will help analyze whether the plan is effective or not.

This does not mean that no further data will be necessary. Additional data must be gathered to ensure the team is on track with the project and to quantify the calculated savings. It's critical to ask if the predictions match the actual results, and make adjustments as the project moves forward.

Conditions change as projects move forward. There may be a need to adjust the timeline or project scope, but the key is to build on the baseline data and not contaminate it with new data. If anything, compare baseline data to current data and evaluate the trajectory of the project. It should be as simple as comparing two points in time, which are the "before and after" conditions.

Here is the key to data collection that isn't pointed out in textbooks. Obviously, there is a need to make an informed decision, but too much data can be a problem. The key is to only collect the information necessary to make a good decision. Remember, other people on the team may have additional information to collect. Make sure to get all of the required information, but keep it to a minimum for ease of analysis.

Collect enough information to get a good snapshot, use the proper Lean/CI tools to evaluate the data, and implement the most effective items by priority. Save enough paperwork and data to tell the story. Anything more than that is unnecessary.

Example 2

Here is another example of capturing too much information to accomplish something simple. Forming metal is pretty common in the industry; however, there are two trains of thought to consider outside of full automation. The first thought is controlling every parameter on every setup screen and holding these parameters during production. The second thought is most of that stuff can be considered reference only, and there are only a few items on each screen necessary to successfully do the same thing faster.

Look at it from a commonsense standpoint, and this should be easy. Consider there are 25 parameters per screen, and there are five setup screens. Production runs on a separate screen, and there are twenty items to monitor. In total, an operator would have the tedious responsibility to monitor and control 145 parameters for setup and production.

How many of the 145 parameters contribute to successfully *forming* a finished part? The correct answer is all of them. How many of the 145 parameters contribute to making a *dimensionally acceptable* part? Suppose the correct answer is five, and it came from an operator with thirty years of experience. Allow me to explain the difference.

The project we were working on was focused on setup reduction. My supervisor demanded that operators save everything from every screen, and he was not settling for less. The operators complained, but the directive was to detail every parameter—no questions asked. It didn't take long for the setup reduction project to go grossly off-track, and the results were obvious. The new setup time was twice as long as the original time, and operators were upset because my boss wasn't listening.

Here is the flip side of the story. I spent time in the department after the event and found that there were only a few things to monitor to make a *dimensionally acceptable* part. The operators and I tested the theory during production, and it was successful. Some of the setups took eight hours or more for the most difficult parts. They were the parts that we focused on because eight hours was completely unacceptable. Especially when you run small quantities of product. The truth of the matter is setups and changeovers generally took longer than the production run, which is not a good situation to be in.

Moving on, there were six forming operations in the sequence required to make an acceptable part. We focused on hitting the critical dimensions and where we had to set the machine to start production. It came down to monitoring the position of the back gauge and forming each bend properly. The point was to simplify things and make the setup process faster, not dictate ridiculous requirements that add absolutely no value.

Another aspect of forming metal is consideration of the tolerances. It all starts with the size of the blank and how it compares to the dimensions on the print. It is necessary to know this in order to ensure all forming operations can be completed. In short, metal stretches when it is formed, so this must be taken into consideration when designing a blank. If the blank size is right and the forming is done per print, a dimensionally correct piece will come off the last press.

The operators and I set the parts up to meet the most critical dimensions and tolerances first. We adjusted the first and last operation to distribute the excess metal evenly whenever possible. The setup dimensions that we came up with were saved for the next production run, so operators were always ahead of their setups. Once the team had a good part, production could continue until the short run was over.

The results of this activity drove the setup time for this particular family of parts from eight hours down to two hours. We also standardized the setup across all shifts to ensure everyone was on the same page. The key to success was spending time in the area, listening to the subject experts, and implementing something that made their life much easier.

The operators and I continued to work together in the same fashion, and we were able to reduce setup times for many other difficult parts. Overall, the area realized a 35 percent overall reduction in setup time and sustained the improvement until I was laid off a year later.

What does that do with respect to head count? Do we lay a few people off and discourage everyone that worked hard to make this happen? The answer to both of these questions is we increased capacity through the setup reduction and improved quality and scrap numbers significantly. We were able to absorb more work in-house and flex our schedule as demand changed.

We didn't need to lay anyone off because we were able to run more products to support sales. The scrap reduction was caused by using the new setup method. The operators were able to reduce setup pieces from an average of six per part to an average of two per part.

Either way we look at the results, I guess there is a win involved. The difference is I am extremely proud of my track record over the course of my career. I have worked on dozens of improvement projects, and *nobody* has been eliminated. My track record also proves that everything mentioned in this writing works. The proof is in the results, so there's no need for me to argue.

The purpose for sharing my stories is not to eliminate the Lean/CI practitioner that continues to teach at all levels of the methodology. There is a definite need for experts, but there is a greater need for standards and basics at the ground level. One of the people that trained me was trained by a sensei at the Toyota Academy in Japan and holds a master black belt in Lean Six Sigma. The man is a Lean/CI walking dictionary, but the thing that makes him so awesome is he is very approachable.

One piece of advice he gave me was to never forget to engage and teach others at the same time. People are what makes it work. He would teach me anything I wanted as long as I held on to showing others the craft. Honestly, I have no desire to be the number one expert on Lean/CI; however, I have a great desire to make sure our nation can apply the basic principles no matter where they work or live.

Lean/CI in the Classroom

Consider the idea that our kids graduate with a STEM style of learning that includes the basics of Lean/CI. My opinion is workplace organization, 5S, and understanding process flow lends itself well to a basic understanding of Lean/CI. Additionally, I believe there are many advantages to teaching Lean/CI fundamentals in high school. The overall advantage is Lean/CI fundamentals apply to all processes and workplace standards.

The first advantage to early education is no matter where graduates work, the basic understanding of terms and principles is already in place. I can put people through a one- or two-hour refresher course, including any differences within a particular company. It will standardize the basic teachings of the main tools that allow Lean/CI to succeed.

Another advantage is it lends well to supporting our economy at this critical time. Consider the additional labor gained by each company by not having to teach basic stuff to everybody. Employees can hit the ground running faster and do not have to go through the same training whenever they change jobs. Many programs I have developed for companies are one to two days, and they cover the exact same information. The only difference is the information is written or presented in a different way. Seriously, think about the advantages.

Many managers and Lean/CI practitioners get highly offended when I say these things because they have wordsmithed their way into a career. The sad thing is I have lost all respect for them, which is why I focus on people.

There is an absolute need for people to teach the higher-level techniques of Lean/CI. These are the people that train the trainers, and they will drive the larger improvements. Lean/CI should be treated as an improvement tool for each individual, company, and nation. It shouldn't be treated like rocket science because the whole world is practicing Lean/CI. Knowledge is great, but application is where the rubber meets the road.

The underlying tone of this text has been redeploying assets to add value in other areas. The exact principle applies here if we move some Lean/CI training to the K–12 classroom. The idea is to create a competitive advantage for each company by teaching specific advanced skills when necessary and how they apply to that company.

The difference is the ability to gain an advantage over competitors. It also contributes to additional profit by improving delivery, quality, and throughput. Finally, there is no need to recreate something that is already a global standard. Build upon the basics, and create success through people and their ideas.

Pushback? No doubt there will be pushback in several areas, and everything needs to be worked out. Who will create the curriculum for the classroom? Who will teach the material since we already have so many experts? Am I trying to undermine Lean/CI through discrediting the people we have in place?

First and foremost, the curriculum has been created by a gazillion people and countless software companies. Most engineers I know have a copy of this type of training in their possession. Honestly, this is not a good thing, but it's true. I already said who developed the methodology, so that argument is over. Standardization of the basic fundamentals on a national level needs to be the first step.

The key to success is to bring the right people to the table to establish the proper curriculum. Global companies set global standards, so experts in Lean/CI must be involved to cover the proper information and to develop the standards of application after high school graduation. It's a matter of alignment, and the need to eliminate anyone is absurd.

As I said, most of the curriculum already exists. The depth of application in each company will depend on the ability to apply commonsense principles without investing a lot of time and money. My point throughout this text is there are savings to be had if we use the proper tools at the proper time. Additionally, the ability to provide smaller companies with an advantage lends credibility to this idea.

Any and every improvement within the overall process can mean the difference between winning a contract and laying people off. Smaller companies need every advantage they can get to be competitive. It seems logical that alignment of Lean/CI training would contribute to their competitiveness.

Hopefully, I have made my point about not eliminating anyone. Lean/CI is critical at every level of its application, and the more we train, the better we will be. I am well trained by very credible people, but I tend to stick to the basics and engage people. We get a lot done without fancy presentations and unnecessary paperwork. Maybe that makes me a blue-collar Lean/CI practitioner, but I am good with that determination.

Short Story

Allow me to share my first encounter as a young engineer. My responsibility was to investigate issues in manufacturing and suggest improvements to the product improvement team. Actually, it was no different than what we call continuous improvement today.

Anyway, there was a problem with a dimension on a part we manufactured in-house. I had to go to the manufacturing floor to talk to the operator, and the following is my recollection of what happened.

I gathered all of the necessary information such as drawings and the actual request for improvement. Again, this was the first time that I had to talk to anyone outside of the engineering department.

I was a little apprehensive as I went to the machining center where the guy worked. I arrived and found a six-foot-plus muscle-bound gorilla. He really was a big dude, and he was intimidating as hell to walk up to and start a conversation with. It didn't help matters that engineers were required to wear a tie as part of the dress code.

He stopped working and turned slowly toward me. There was no smile, no nothing! It was obvious that I was interrupting his work. He walked up to me and said, "Let's get one thing straight. I would rather choke you with that tie than talk to another engineer. What do you want?"

What do you say to someone like this? This may not end well, and I was taken aback. After a short pause, I replied, "I don't exactly like wearing this tie, and I really don't feel like being choked today." He had a puzzled look on his face, and now he was taken aback. And then came the awkward pause where both of us were trying to figure out what to say next.

I stepped up and began to explain the change request. Then I asked what he thought we should do. His response was classic: "Are you asking me how to fix this? You're the engineer. You tell me." Here we go again! I laughed nervously and humbly admitted I was new at this and I didn't really know how to run his machine.

He stepped back for a second and looked at me in a weird way. Then he stepped toward me and explained how the machine functioned. He also explained why there was no way to create the feature on the print. Finally, he suggested what he knew would work to correct the situation. It's funny how my tie seemed to fade away after I showed some respect for this man and his chosen craft.

Here is where it all starts. Historically, the man had been *told* what he had to do to fix something for his entire career. Even if he disagreed, he still had to follow the wrong process. Actually, this is the reason I went to the floor in the first place. An engineer made a change to a drawing that could not be replicated in a machining center. It worked on paper, but the part could not be made to print in this situation.

It was the first time I went to the manufacturing floor, but this is the one interaction that would shape the way I needed to approach every change or improvement. Show respect, engage people, and implement the best practice. It is thirty-five years later, and it has been the best thing I ever did.

After the operator and I were finished, he shook my hand and told me how impressed he was that I would listen. It was the first time in his tenure as a machinist. I think it's only fair that I thank him in return because he had the courage to challenge the norm, not the engineer. It has made all the difference between success and failure for me.

Easy Money

Here is where the rubber hits the road when it comes to smaller companies being more competitive, and larger companies having better standards and processes. It doesn't take a rocket scientist to find savings in almost every process, including the assembly of a PB&J sandwich.

The best way to describe this thinking is with the first question I ask when I enter a department. What is your biggest problem? Ninety-nine percent of the time, someone will speak up. Once a person speaks up, the rest of the department tunes in because they want to know two things. Is the Lean/CI practitioner going to listen to them and help come up with the best solution? How arrogant is he/she? Think back to the examples, and this should make sense.

This is the most critical moment in the whole process, so think about the things you would say and the way you expect people to treat you. Back to the original question. What is your biggest problem? Once you get the short list from the people involved, create a plan of attack to help them help you and help the company. This is the true definition of teamwork!

As I mentioned before, there are many small things that can be accomplished with little or no cost. It's these small things that help generate creative thinking, new ideas, and improvements because people see that their input helped. They also see that the Lean/CI person listened to them, so they will be more willing to participate. Finally, have fun in the name of improvement, and celebrate small wins. Nobody ever told me this couldn't be fun!

There also doesn't need to be a ton of paperwork to satisfy someone's craving for information. You do not need to collect mountains of data every day to know where the opportunities lie and if the improvement is working. Most importantly, engage and encourage people to get out of the box and look at the situation as a team. People are the key to getting things done.

Common sense will satisfy the majority of the required homework necessary to make improvements. Use data to prove the root cause of the problem and to prove that the correction or improvement worked. If there is an additional need to store the information, then create a database. Keep the database nomenclature rather simple even if the information is extremely technical. It will serve you well in the long run.

Example 3

I want to change gears and share some of the simpler improvements that made a huge impact with little investment. Reorganizing a machining center, changing product flow, and revising methods of delivery can contribute to quick savings and bottleneck relief. Standard work can improve processes by introducing best practices to ensure the same outcome for each item produced.

Earlier in the text, I referred to a project that relocated a department from one facility to another. The entire facility was new, and there were several other departments moving as well. We already knew about the improvement in component travel, but we felt there was more we could do. The mantra of our team was "If you're going to move it, then improve it." This is an example of small wins adding up and taking a good project to another level.

The first thing on the agenda was how to set up the new area. Somehow, it is easier for me to look at the goal and work backward from there. We already know what we have. We know we need to get rid of stuff, but we aren't sure what yet. What about component delivery? What needs to happen to get to the goal?

Currently, there are storage racks full of random parts that belong to any one of several custom trucks being built. The storage racks are located on the other end of the current building, and every employee retrieves their own parts. The goal quickly became improving the storage and delivery of parts. It would be critical to functioning more efficiently in the new location.

The last item we addressed was the placement of everything in the build bays. We identified several items such as standard tools and worktable layout. There were electrical drops and appropriate weld gases in place. We even had a place for the garbage cans in the layout. The trick to success was to ensure the new layout worked for everyone involved because they were the people that had to use the space.

Working backward, the layout of each assembly bay was complete, and it was time to look at component storage. What can we do to improve in this area? Automotive companies use a "water spider" to deliver parts to their assembly lines. One minute of downtime on an automotive assembly line can cost millions of dollars. The very last thing they want is to have a line go down for lack of parts.

There were several advantages to adopting this method of delivery in the new facility. We created a drop zone for incoming parts and separated them by the truck they belonged to. We placed racks in each bay to create point-of-use storage. The advantage is employees could retrieve their parts without leaving their work area. In short, the change led to better organization of our work area and a safety improvement through removing clutter. Another safety concern was eliminated through reduced traffic in the aisles.

It took one person to do all of this as opposed to four out of fifty employees being out of the department at any given time. The cost savings are considerable if you look at direct labor within the measurement of efficiency. Four people at eight direct labor hours each day compared to one indirect labor person at eight hours a day. It isn't voodoo math. It is real savings from my point of view.

The calculation is clear no matter what your thinking is. The obvious gain is three people adding value instead of random employees looking for parts during the complete shift. The question becomes whether or not to lay off the three people you just gained by changing the delivery of components.

Again, here is where the argument begins. Many managers feel three people must be laid off in order to realize the gain. You can do that, or put three people back in the department. There is an additional option to move the people to a new area. The point is to utilize the labor improvement to increase efficiency in the department or fix a bottleneck somewhere else.

LEAN MANUFACTURING

Our team was able to deliver more products with zero effect on quality, and our sales increased because we were able to significantly shorten the lead time for delivery. The result was a 12 percent sustained improvement in efficiency over two years. For the record, the project was blessed and verified by the director of finance.

Here are the simple calculations:

> 3 people × 8 hours × $25 per hour = $600 per day × 5 days = $3,000 per week.
>
> The total is $3,000 × 52 weeks = $156,000 one-time savings.

or

> 3 people × 8 hours = 24 additional assembly hours per day.
>
> 24 hours per day × 5 days = 120 additional assembly hours per week.
>
> The cost of labor doesn't change if you move people.
>
> Current sales = 400 custom trucks annually. The average cost per truck is $250,000.
>
> Sustained 12 percent improvement = an additional 48 truck capacity per year.
>
> 48 × $250,000 = $12,000,000 in sales due to the 120 hours of additional labor per week, and no change to current labor costs. The savings also carry over from year to year. The decision is yours.

With respect to reorganizing a machining center or improving product flow, look at the overall layout and how the area fits into the bigger picture. Drop zones are simple to create, and flow is not much more than common sense. Ask where the product is coming from and where it is going when it's finished. Many Lean/CI practitioners refer to this as upstream or downstream.

Think about it like this. We didn't reduce head count, but we gained value through improving the flow of our product through the factory. It also contributes to success by raising the morale of employees. The cool thing is companies can do things like this every day because the immediate payback outweighs the cost.

My philosophy has always been that small wins add up. The best way to find the small wins is to work with people. Ask them what they think, and you will be surprised what comes back. Remember, a large project doesn't have to be much more than an accumulation of small wins organized to get something done more efficiently.

Timing Is Everything

For those who aren't clear about the role of the water spider, here is a brief description of our thinking during the previous project. The design of each work center was done, so the next task was to get rid of bottlenecks. The largest bottleneck was in the storage and delivery of parts. Delivery routes and timing of rounds had to be established for a material handler to follow to deliver products to the areas necessary.

The first task for our team was to establish the best time to complete each round of delivery. We looked at it as if a doctor is making rounds in a hospital. Fabricated products arrived randomly during the day, and quantities varied. This is where the drop zone became important. The contents of the drop zone would be sorted between rounds of delivery and delivered to their proper location/truck as required.

Purchased products arrived every morning at eight, and were ready to deliver to their required location by ten. The whole process was very easy because we knew what had to be delivered and when everything was available. We also knew where everything was going. Once we were able to determine when and where, we were able to continuously deliver incoming parts to their proper location. Think of this similar to the way just-in-time delivery works. Right parts. Right place. Right time. And in the required quantity.

Standard work and role expectations can be a good thing if done properly. I lost track of the number of standard work instructions that I have written. Many companies create a format/template that everyone must follow. This is a good thing because the necessary information for any job can be found in the same location.

Expectations can become a touchy subject at times, so it is important to understand the principles of engaging people. The work instructions that are created must serve the employees, or the purpose is completely lost. That being said, every instruction must include input from the people that perform the work. If you allow people to set the expectation, they will monitor for compliance.

The statement I make to operators is I'm not here to tell you how to do your job. I clearly let them know that I am there to document how they do their job, including any suggested improvements. It's a matter of establishing the best practice of performing a series of tasks.

The benefits are obvious to places that run more than one shift. Creating standard documents guides the company (and coworkers) through training, performing tasks, and accountability for following instructions. Quality improves as the amount of variation is reduced, and the output from the area becomes more predictable.

Feedback from management and operators has been very positive, and sustaining the quality of the product has fundamentally improved in every case. The other effect of standard work is less complaining about mistakes or laziness. Accountability is built in with the time it should take to complete a process, which has been established by a team including subject experts and coworkers.

Like it or not, many people watch other people, and they will pace themselves with the slowest person. The reason is that person is getting away with it, so why not do it too. Standard work helps keep people on task, improves quality, and provides a tool to set the expectation of performing an operation. Accountability from coworkers and supervision will naturally support the expectation because people do not want to look bad.

On a totally different front, there is also a critical timing element to preventive maintenance. Systematic maintenance and repairs are necessary for every piece of equipment involved in the production process. It shouldn't matter if the machine is meant to support production or run production; unplanned downtime is very costly.

We will move into the thought process in the next section, but keep in mind that machines must run in order for a company to survive. The timing of maintenance will make the difference when it comes to keeping the machines functioning properly. The priority of maintenance items depends on how critical a failure will be when it happens.

The trick to success with maintenance timing is to get the most out of every planned shutdown. Start by driving to eliminate the expensive unplanned downtime and keep money within the company. Use the savings that are generated to invest in replacement parts, improve throughput, rearrange areas, and create value for all stakeholders. This can become a vicious cycle of improvement, so be prepared to think differently. Start doing the obvious things that add immediate value to your operation.

Reliability and Availability

Machine reliability and availability have always been important in manufacturing. The truth is you cannot produce parts if the machines aren't running. If the machines aren't running properly, there may be an obvious quality issue that arises. The part that baffles me is the number of arguments I have had to fix a machine properly compared to running production until the machine breaks down.

Over the course of my career, it has become clear that assets seem to be purchased with the thought that they will run forever without maintenance. The company has a 24-7 operation, and downtime is money. However, the norm is to react when machines fail instead of maintaining them properly. Allow me to prove that this thinking actually costs more than proper maintenance. If you are waiting for a machine to fail, you are already too late.

Clearly, unplanned downtime is the enemy of production. There should be as much focus on preventive maintenance as there is on production. The reason is having the ability to remove unplanned failures and replace them with planned activity and production.

There are two aspects to this. The first is an assessment of the recommended intervals from the manufacturer. Do they line up with the historical failure of components? Is there something in our process that contributes to premature failures? The important thing here is to ask the right questions to get to the root cause and solve the problem.

The second aspect is to shorten any unplanned downtime to a minimum and get back to production. Obviously, Pareto charts and reasons for downtime need to be generated to resolve the biggest issues first. There are other tools such as five whys and A3 formats that formalize the process of problem-solving. Use all of the available Lean/CI tools to your advantage anytime you can. The key to success is getting in front of the maintenance schedule in order to maximize machine availability.

Equipment manufacturers provide manuals that contain all of the information necessary for proper machine maintenance. There are replacement parts, average life expectancy, and intervals when maintenance is required. Even though the information is readily available, many companies fail to utilize it to keep their machines running.

The key to success is to transfer or use the information to set up maintenance intervals. Make sure replacement parts are available and interruption to production is minimized when a machine is brought down for maintenance. Create a crash cart or kit parts as necessary to speed up any maintenance work. The best way to describe taking a machine down is to liken it to a NASCAR pit stop.

In the early days of NASCAR, a pit stop might take two minutes or more. Generally, this is the time it took to change four tires and fuel the car. Today, a fifteen-second pit stop is the maximum time it should take to change four tires and fuel a car. Here is where the small wins can help a process, because each time the machine is brought down, the opportunity for improvement presents itself.

NASCAR pit stops are choreographed to the step. Everyone has a specific job and the time necessary to complete the task. If there is a way to improve, the team makes the decision and implements the change. Does this sound familiar? It should, because it is exactly the way Lean/CI is designed to work.

In a perfect world, machines run without failure, and 5–10 percent of the available time is necessary to complete required maintenance. The machine is brought down, repaired, and returned to production just like a pit stop. Wouldn't that be nice? Is it even possible? The answer to both questions is yes, and all it takes common sense to get there.

Outside of this is the ability to troubleshoot issues when unplanned downtime does occur. Honestly, this part is more critical than preventive maintenance, from my experience. Repairs can cost up to ten times more through expediting parts or having a company technician come out on short notice. Additionally, the repair takes longer because the event was unplanned. And don't forget that someone may need to be pulled from doing something else to fix the problem.

Here is the worst-case scenario that should make this point pretty clear. Consider the idea that you have a process that coats parts and the tooling has to be stripped after three uses because of coating buildup. The coating process works well, but the cleaning process for the tooling is failing miserably.

The machine used to clean tooling is a high-pressure water blast unit, and there is no maintenance schedule in place. There are also no written instructions on how to fix the machine, and the maintenance people do not have much experience with this type of equipment. A technician from the manufacturer costs $100 per hour plus expenses, and it takes two days to get here.

Unplanned maintenance was happening daily, and the ability to coat parts was in serious jeopardy. There were very few spare parts in stock, and some of them were obsolete. Here is the point to all of this. The information was available in the user manuals, but nobody took the time to look at the maintenance schedule. Had the person who bought the machine done this up front, the reality is a different method should have been used. Unfortunately, it's too late now, and something must be done.

Another important note is the failure to use the maintenance scheduling software that was provided by the manufacturer. Even though everything had been provided to successfully strip tooling, maintain the machine, and stay ahead of the maintenance schedule, it was blatantly ignored.

Was it because of the greed of management and their desire to deliver to the key measurables? Was it the failure of an engineer to complete the process of bringing the machine up properly? Was it because of poor workmanship in the maintenance department? Was it the fault of the manufacturer? Unfortunately, this case includes a little bit of everything.

First, there is a constant battle to run production or fix the machines. Maintenance gets the blame for machines not working properly, and unplanned downtime is happening far too frequently. The argument for that is you cannot fix the machine properly if you don't schedule maintenance. The intensity of these arguments can be extremely high because there are conflicting interests. It boils down to fixing the machine properly or placing another Band-Aid on the problem.

The truth of the matter is proper maintenance leads to machine availability and reliability. You can adhere to a production plan with confidence because machines are putting out good parts, not scrap and oil puddles. Delivery to customers remains on track, and the potential for new business comes back in return.

Small companies cannot afford to pay ten times the cost of a repair, and they seriously cannot afford to have a machine down. Maintenance information is provided with the purchase of each piece of equipment, so it makes sense to use it. There is no need for an elaborate planning system if you can perform the proper maintenance items when necessary.

If you are part of a large company, it is in your best interest to do the same thing. The important thing for everyone to remember is to stay ahead of the maintenance schedule in order to improve capacity, quality, reliability, and delivery.

Step back and ask yourself where you stand in the maintenance argument. Would you run the machine to failure short-term to have unplanned downtime later? Would you take the machine down, fix it, and reliably get back to production? The answer depends on how you look at the value of the assets.

As a rule, the goal of machine availability should be 90 percent or greater. This feeds directly into world class manufacturing standards and shows customers that you care about the machines that make their product. It doesn't matter if it's a drill press, forming machine, or expensive coating process; customers judge suppliers every time they walk into your plant. Future businesses may depend on their last impression.

Compare this to 30 percent or more unplanned downtime, and it is easy to see why every company should fight tooth and nail to eliminate the problem. Small companies simply cannot afford it, and others have horrific pricing to compensate for the inefficiency in their processes. Get away from unplanned downtime, and focus on keeping machines running good product. The payback is worth far more than the investment.

Running the machine to failure presents a different set of potential outcomes over and above unplanned downtime. There is a possibility that the machine will fail sooner and force you to fix it. Another possibility is the machine may complete the order for the customer, but it experienced additional damage during the process. Add the possibility of low or unacceptable quality, and this idea loses credibility quite quickly.

On the other hand, the proper fix will take one shift as soon as we receive replacement parts, which will be here by the end of the current shift. One of our maintenance technicians from the first shift will stay over to speed up repair and bring the machine back up for production. The difference between the two scenarios is delivery of a questionable product today or delivery of a good product tomorrow. Again, the choice is yours to make.

The end result of the second scenario will return the machine to proper function and prolong maintenance until the next planned repair. The negative effect is someone has to call the customer to explain the situation and tell them their parts will be delayed one day. It seems like a small price to pay for an unplanned failure; however, the goal is to eliminate them to begin with.

Large assets are in place to produce large quantities of whatever, and naturally, management wants all they can get. The problem is assets do not produce anything when they are idle. The goal here is to put common sense into the equation and share that a good plan beats no plan any day of the week. As the saying goes, "If you fail to plan, you plan to fail."

Large companies may have some ability to absorb the cost of bad planning, but smaller companies may not fare as well. Small companies depend on assets and need to have the best plan possible to keep their machines running. The cost of downtime can be expensive, but it's cheaper than running to fail accompanied by unplanned expenses.

My final point is all companies can benefit from reliability and availability simply because of the fact that you can deliver more products in less lead time with better quality than the competitor. Contracts are awarded to one company over another because they can prove there is a maintenance plan in place. Do not mistake the importance of keeping equipment operational.

It is critical that everyone understands that preventive and planned maintenance does not have to cost a lot of money. Success lies in the ability to identify the critical items on the machine, monitor them for replacement, and make a plan to repair the machine before it fails. Evaluate the unplanned failures and revise maintenance intervals or drive deeper to a root cause when necessary.

Allow me to add that unplanned failures must be captured and driven to the root cause whenever they occur. Here, again, a mountain of information isn't required. Use the available Lean/CI tools to find answers, and do what is necessary to eliminate the problem from happening again. My best advice is to use common sense as a guide to continue to do the right thing as you progress.

How difficult is it to set up a plan for a maintenance schedule? It can be extremely difficult depending on the asset, but the key is to keep things as simple as possible. Usually, the machine manual contains everything necessary, and there is already a planning matrix for repairs. Use this matrix and all other available information to get started.

Keep track of the unplanned failures and apply the proper fix when they happen. Feed the information into the current maintenance system to determine if anything needs to change. The key is to stay in front of the repairs and plan when the machine is going to be repaired. If you fail to do this, the machine will dictate when repairs need to happen.

Let's circle back to the water blast machine used as an example. The root cause of all of the failures ended up being the blast water was too dirty. The dirty water ate away all of the seals and spray orifices in the entire machine. Motors that were designed for this environment were failing because of the contamination in the water. Planned maintenance was never done because the demand for the machine was so high.

Evaluation of the root cause fell in my lap, and it didn't take long to figure out what was happening. The problem compounded because once we fixed the cleanliness of the water, the machine would need to systematically be repaired and brought back to proper operation.

The original cost of the machine was just over a million dollars, and the projected repairs were damn close to the same number. All of this was the result of having no plan for maintenance of the machine. The information was available, but it was ignored. Many of the internal alarms were also ignored or overridden, so the machine deteriorated quickly simply because nobody took ownership.

There is another small lesson that falls inside this project. As you know, I am a strong advocate for bringing the right people to the table to resolve issues. As this catastrophe unfolded, we still hadn't found an efficient way to clean the water. Water filtration machines that could support our requirements were about $50k, but we were told there was no money in the budget. So basically, we had to make a silk purse from a sow's ear, and the situation was hopeless at best.

A discussion between a facility engineer, two millwrights, a senior operator, and myself came up with an alternative solution. Somehow, our conversation moved toward settling mash in whiskey. Don't tell anyone, but this is where fun can completely change the outcome of an issue. We had the same issue as whiskey makers, and now we had the same idea! Why not set up a series of settling tanks? We did, and the rest is history.

We designed a set of barrels that would settle most of the large particles in the water and allow the filters inside the machine to do their job. It was hillbilly-looking, but it worked like a charm. The final cost of the cart was about $1,500, and the four of us changed the condition of the water enough to get the machine running better than ever.

The settling carts remain in the process to this day, and the payback has been the ability to repair the water blast machine and clean tooling for production. The settling tanks are replaced when they fill up and are disposed of according to federal regulations. The only thing to add is we did this as a team and implemented the best possible solution for long-term results. We didn't make moonshine, and it wasn't pretty. But it was effective, and it worked.

Who? What? Where? When? Why?

Machine designs require a bill of materials to show the main assembly, subassemblies, and component parts. The main assembly is an overall picture of the items necessary during the assembly process. Subassemblies are separated to detail the components at a lower level. The component parts pretty much speak for themselves.

Once this is complete, an indented bill of materials is created from the drawings, and the design is released to production. Maintenance schedules can be determined during this process and provided with each product sale. What does this have to do with anything?

It has everything to do with everything! Bill of materials can be extremely difficult to interpret unless you have prolonged exposure. They are confusing and meant for engineering and assembly, not necessarily the maintenance department. Over time, the description that best describes machine maintenance is to liken it to tree roots.

Here is the point. Keep everything simple to understand because it will benefit the masses. The idea of a tree root system can be visualized, so that's why I suggest it. The trunk serves as the main piece of equipment, and the limbs and branches represent the subassemblies and components respectively.

For clarity, a subassembly may be a motor, weldment, or anything consisting of more than one part. The trick is to generate information that fits into this visual because it helps everything fall into place somewhat naturally. Stick to the idea that simpler is better as you work into a maintenance plan.

Use the noun *naming convention* for identifying replacement parts because it makes it easier to find certain things, and it is the acceptable, best practice. A simple example would be a motor replacement. Noun naming suggests using "motor" first and adding additional information as necessary.

The key is thinking ahead will save a lot of time later. If a maintenance supervisor needs a specific motor, he or she would use motor to start the search. If everything is set up correctly, a list of available motors will come back.

The idea of simplicity is made because there are too many Lean/CI practitioners who feel the need for more documentation than an issue is worth. The point is you do not need to capture gnat's ass details to make an impact. Capture the most important aspects, and store the information in a place where people can easily access and use data. Finally, you should not have to spend more time documenting than doing. Spend your time solving the issues, not working on someone else's promotion.

What is really necessary to perform proper maintenance? How much information should you provide skilled trades such as electricians or millwrights? What about smaller companies who have to fix their own machines?

There is a simple answer to these questions, and the following describes one way to look at the situation. There are five basic questions to ask when setting up a maintenance plan. I have used this method at small and large companies, and the results have been very good.

Question 1: What is going to be fixed? Thinking back to the root system, the description of the repair can be as simple as identifying one component or subassembly. This information can be used to get replacement parts, necessary tools, and additional instructions if available.

Question 2: Where is the fix located? Depending on the company, this can mean several things. The description may include plant number, department number, and machine center, which is location-based.

A company may also use a problem-solving method to drive to the root cause for every breakdown, which points at one thing. This is component-based.

They are completely different scenarios, so flexibility and simplicity are beneficial either way. The question is the same either way.

Question 3: Who is going to perform the fix? The answer is easy with respect to large companies because skilled trades make the determination. A work order is written, and eventually, someone will show up and fix the problem. I say this in jest because I have many good friends in the trades.

A small company may have a different scenario than this. Resources are at a minimum, and people who fix the problem may need to be pulled from production. Either way we break it down, the question is still the same.

Question 4: When will the fix take place? If the event is unplanned, obviously, the need is immediate. Fix issues as soon as possible, and analyze what information is currently in place. Improve where necessary. Remember, timing is everything.

One thing to keep in mind is how critical the situation will be if you postpone any repairs. Refer to your Pareto of breakdowns and maintenance, and you will be better prepared to make the right decision.

Question 5: What is necessary to perform the fix? This question came up because some companies have a maintenance shop where the tools are located. Individuals have to walk back and forth to get tools, and that was inefficient at best. If you know what is necessary, you can better prepare.

In addition to this, spare parts, required tools, and instructions for the fix can be included. I referred to this as a crash cart earlier in the text. If there is an opportunity to use kitting for spare parts kits, make sure to eliminate waste by eliminating the unnecessary parts in the kit.

The final question of *why* something broke down is completely separate from the previous questions. This is the fundamental approach of root cause analysis and problem-solving. My advice is to use what works best and continue to strive for perfection. There is no need to make Lean/CI rocket science because it is not necessary.

Task Lists

Last but not least, task lists are the most successful tool I have ever used during a project. When you think of it, Lean/CI boils down to having things that need to be done in the necessary order, by the right people, and at the right time. It doesn't matter if it's a machining center, material delivery, 5S, or maintenance. The fact is tasks need to be completed to make the project a success, and a task list serves as a guide for everyone to follow.

Task lists are excellent for improvement projects for many reasons. Most importantly, the task list represents everything that needs to be completed for the project. It doesn't matter what the size of the project is; there is a need for planning and execution. Some projects include skilled trades, outside contractors, engineering, and any other entity you can think of. The best method to keep all parties on track is to introduce a task list.

It becomes critical that every individual on the project knows what is going on for progress as well as safety. What is done? What is next? Who does this? Who does that? Are the employees safe? It doesn't take long to find ten people standing around with no clue about what is happening. This is the primary reason for developing the task list and keeping it up to date.

Every project has deliverables, so set up the task list to support them. For example, if your goal is setup reduction, list the tasks necessary to get there. Accountability can be built in with assignments, priorities, timelines, and completion dates. Keep it simple, but include the necessary information to support each task and the overall project. Hold people accountable for their given tasks, but be prepared to offer support or change some things as the project moves forward. If all goes well, the timeline will work itself out. Flexibility is the key to success.

Another example of a task list may include changing a pump on a machine. There are several steps necessary to complete the task, and they must be done in a certain order. This is a small part of a larger project, and it's critical that there aren't any mistakes. Once the pump is replaced, there are other tasks that need to be completed subsequently to finish the project.

An important element to the task list is the daily review of the list. This will help keep the project on track and keep everyone up to date. With respect to larger projects, it may be necessary to review the list at shift change, morning meetings, meetings after lunch, and meetings at the end of the day. The point is communication is another key to success, and the task list is the best visual tool that one can use.

The contents of a task list can vary wildly depending on the scope of a project and/or the size of the company. Many smaller places may choose to use a simple list that is checked off as things are completed. The list can be written on a napkin or scribbled on a piece of cardboard as far as I am concerned. Seriously, it doesn't matter as long as it helps keep people on track.

Some of the most important items to identify for each task include the "who, what, where, when, why" concept of maintenance. Other items may include work order numbers, priority levels for completion, replacement part numbers, and anything else that should be tracked. I do not want to create an all-inclusive list because it is only necessary to track the most important items as they apply to your situation.

Many of the task lists that I have used are printed on a twenty-four-by-thirty-six-inch piece of paper and hung adjacent to where a project is taking place. There are several columns of information necessary to ensure the timing and execution of the project tasks. At times, there are several pages. It can be intimidating to walk up to, but the truth is 90 percent of the information for the project is contained within.

This is the main reason I push back when asked to document more than necessary. I already have the information for the project centralized and saved electronically. I also have a database that houses every project we have completed to date. If someone needs to know anything regarding the project, it's already in one spot.

The point of displaying the task list close to the project has a few advantages. First and foremost, anyone from the floor to upper management can walk up and understand everything that is complete, who is doing what, and what needs to be done next. Additionally, everyone involved in the project can walk up and learn the same thing. Finally, it is critical that everyone involved has the permission/authority to note anything necessary to support communication on the project.

It is difficult for many people to speak in front of others, so I use this as an option for them to have a say. The best projects I ever ran worked like this, and I was amazed at the positive reaction. Millwrights and electricians would sign off on work orders and update the task list. If a shift change came, the priority of the work orders supported what had to be done next.

The electronic task list should be updated daily by the project leader, with completed items and any additional notes that were made. Post a clean task list every day to eliminate any confusion. The point is the task list should look professional and not like graffiti on the subway. Simplicity is the key to success as well as allowing others to write on the task list.

There is one caution to using a task list for large projects. The task list may grow to a size that is not manageable in one file. Additionally, it is very easy to keep building items into the file until there is simply too much information to digest. Here, again, I have to refer to keeping things simple. Save the necessary information to tell the story of the "before and after," but do not keep unnecessary details.

Final Thoughts

Hopefully, I have been able to make the case for commonsense Lean/CI thinking. Smaller companies cannot afford to spend money on things that add no value. Large companies may be able to absorb the cost, but do they really capture the value of each improvement? The difference may mean the survival of your company.

One of the goals was to show the reader real calculations that add up to real money. I supplied three examples and included the basic information to set them up. I also shared the simple calculations that were used to make critical decisions along the way. These decisions kept the team on track and always contributed to the overall goal.

The question about laying people off versus redeployment of resources is meant to make you think differently about savings. The additional goal is helping small companies grow with improvements without spending a ton of money. There is no voodoo math involved here. It's basic math that can be applied in a wide variety of ways.

Another goal of this writing was to encourage the engagement of people and discourage the alpha-male thinking that has become obsolete. It isn't worth my time to argue about semantics when the proof lies in the actions of the team.

Work with people, and engage them to get ideas. Treat everyone with respect, and ask for input. Every person at the table is an expert, and their responsibility is to help the team. Start there, and build the team as individual strengths develop. It may evolve into better team performance and successful project completion.

Lean/CI practitioners that want to be successful need to teach the necessary tools for everyone to succeed. Leave your ego at the door because people generally don't care to help when there is no benefit to them. It also upsets people when they are overwhelmed with information, so move to the manufacturing floor to support what was taught in the classroom.

Most of the negative comments within this document were shared by members of teams I have worked with. I have developed a few beliefs as well. The point is we were outranked in every situation, so all of us had to endure the poor behavior of a person who had no clue what they were doing.

Lean/CI is all about the engagement of people, and success is driven by their involvement. My hope is I have made my point with respect to that. Additionally, companies should evaluate the effectiveness of their Lean/CI team by anything but head count.

Another aspect of this writing was to introduce the thought of teaching basic Lean/CI skills at the high school level. It only makes sense to do this because our kids will graduate with the basic skills of Lean/CI to enter any workplace. This also contributes to the competitiveness of smaller companies in a global setting.

All companies can build on the basics sooner and not waste time and money training with the same information as everyone else. Instead, there is the ability to train more people in the Lean/CI practices that create a competitive edge.

My thought is it can't hurt to think differently as the world changes. The metric system freaked everyone out. The truth is I work back and forth between the metric and the standard inch system every day, all day. Get out of the box, and look at the bottom line differently to help yourself become more competitive in any situation.

About the book

About the Author

Randall L. Kies II is a career engineer in manufacturing, Lean Six Sigma, and continuous improvement. He earned his bachelor's degree in science and industrial management from Baker College in Muskegon, Michigan. He earned his master's degree in science and engineering management from Milwaukee School of Engineering in Milwaukee, Wisconsin. He resides in Muskegon, Michigan, with his wife, Angelina, where they enjoy watching the beautiful Michigan sunsets.

www.ingramcontent.com/pod-product-compliance
Lightning Source LLC
Chambersburg PA
CBHW021027180526
45163CB00005B/2145